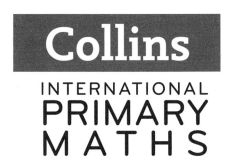

Collins

INTERNATIONAL PRIMARY MATHS

T017343T

Student's Book 2

William Collins' dream of knowledge for all began with the publication of his first book in 1819. A self-educated mill worker, he not only enriched millions of lives, but also founded a flourishing publishing house. Today, staying true to this spirit, Collins books are packed with inspiration, innovation and practical expertise. They place you at the centre of a world of possibility and give you exactly what you need to explore it.

Collins. Freedom to teach.

Published by Collins
An imprint of HarperCollins*Publishers*
The News Building
1 London Bridge Street
London
SE1 9GF

HarperCollins*Publishers*
Macken House,
39/40 Mayor Street Upper,
Dublin 1, D01 C9W8,
Ireland

Browse the complete Collins catalogue at www.collins.co.uk

British Library Cataloguing-in-Publication Data
A catalogue record for this publication is available from the British Library.

Author: Lisa Jarmin
Series editor: Peter Clarke
Publisher: Elaine Higgleton
Product developer: Holly Woolnough
Project manager: Mike Harman (Life Lines Editorial Services)
Development editor: Joan Miller
Copyeditor: Catherine Dakin
Proofreader: Tanya Solomons
Cover designer: Gordon MacGilp
Cover illustrator: Ann Paganuzzi
Typesetter: QBS Learning
Illustrators: Ann Paganuzzi and QBS Learning
Production controller: Lyndsey Rogers
Printed and bound by Grafica Veneta S. P. A.

With thanks to the following teachers and schools for reviewing materials in development: Antara Banerjee, Calcutta International School; Hawar International School; Melissa Brobst, International School of Budapest; Rafaella Alexandrou, Pascal Primary Lefkosia; Maria Biglikoudi, Georgia Keravnou, Sotiria Leonidou and Niki Tzorzis, Pascal Primary School Lemessos; Taman Rama Intercultural School, Bali.

MIX
Paper from responsible sources
FSC™ C007454

This book is produced from independently certified FSC™ paper to ensure responsible forest management.

For more information visit: **www.harpercollins.co.uk/green**

The publishers gratefully acknowledge the permission granted to reproduce the copyright material in this book. Every effort has been made to trace copyright holders and to obtain their permission for the use of copyright material. The publishers will gladly receive any information enabling them to rectify any error or omission at the first opportunity.

Cambridge International copyright material in this publication is reproduced under licence and remains the intellectual property of Cambridge Assessment International Education

Contents

Number

Geometry and Measure

Statistics and Probability

How to use this book

This book is used at the start of a lesson when your teacher is explaining the mathematical ideas to the class.

- An **objective** explains what you should know, or be able to do, by the end of the lesson.

Key words

- The **key words** to use during the lesson are shown. It's important that you understand what each of these words mean.

Let's learn

This part of the Student's Book page **teaches** you the main mathematical ideas of the lesson. It might include pictures or diagrams to help you **learn**.

Guided practice

Guided practice helps you to answer the questions in the Workbook. Your teacher will talk to you about this question so that you can work by yourself on the Workbook page.

HINT

Use the page in the Student's Book to help you answer the questions on the Workbook page.

 Thinking and Working Mathematically (TWM) involves thinking about the mathematics you are doing to gain a deeper understanding of an idea, and to make connections with other ideas. The TWM Star at the back of this book explains the eight different ways of working that make up TWM.

At the back of the book

Number

Lesson 1: **Counting in ones**

• Count objects in 1s

Let's learn

Guided practice

Cross out each sweet as you count it.

How many are there? | 28 |

Lesson 2: **Recognising patterns**

Key words
- **pattern**
- **tens frame**

• Recognise up to 10 objects in unfamiliar patterns without counting

Let's learn

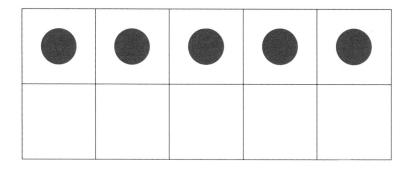

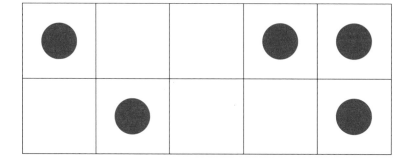

Guided practice

Without counting, write how many dots are in the tens frame.

9

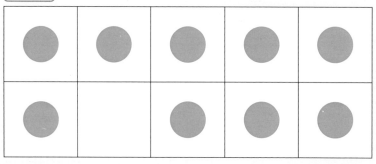

Lesson 3: **Counting in fives**

- Count on and back in 5s
- Count objects by making groups of 5

Let's learn

Guided practice

Draw a ring around each group of 5 flowers. Then count the flowers in 5s.

55

Number

Lesson 4: **Counting in tens**

- Count on and back in 10s
- Count objects by making groups of 10

Key words
- **count on**
- **count back**
- **tens**
- **groups**

Let's learn

Guided practice

Draw a ring around each group of 10 beads. Then count the beads in 10s.

40

Number

Lesson 1: **Counting in twos**

- Count on and back in 2s
- Count objects in groups of 2

Key words
- **count on**
- **count back**
- **twos**
- **groups**

Let's learn

Guided practice

Draw a ring around each group of 2 dolls. Then count the dolls in 2s.

18

10

Lesson 2: **Even and odd numbers**

Key words
- **even**
- **odd**
- **share**
- **equal**

- Recognise even and odd numbers to 100

Let's learn

Guided practice

Draw a line to join each boat to the odd or even bank.

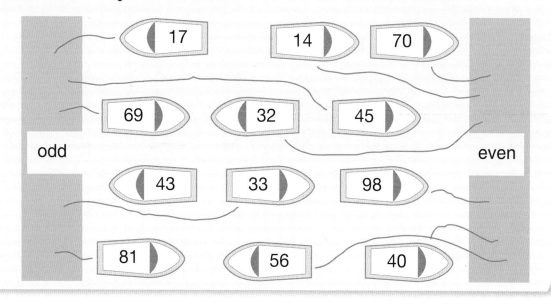

odd   even

17 14 70
69 32 45
43 33 98
81 56 40

Lesson 3: **Counting on and back**

Number

- Count on and back in steps of 1, 2, 5 and 10

Key words
- count on
- count back
- ones
- twos
- fives
- tens
- groups

Let's learn

14, 16, 18, 20, 22, ...

25, 30, 35, 40, 45, ...

42, 52, 62, 72, ...

86, 87, 88, 89, ...

Guided practice

Continue the number pattern.

65 70 75 80 85 90 95 100

Lesson 4: **Estimating**

- Estimate how many objects in a set of 20–100

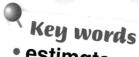
Key words
- **estimate**
- **count**
- **tens**
- **fives**
- **twos**
- **ones**

Let's learn

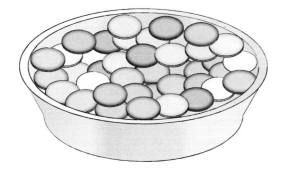

20?

50?

Guided practice

Draw a ring around the estimate that you think is closest to the number of stars.

(20) 50   70   100

Lesson 1: **Counting to 100**

- Count on in 1s from 0 to 100
- Count back in 1s from 100 to 0

Key words
- count on
- count back

Let's learn

Guided practice

Draw a line from 75 and count **on**.

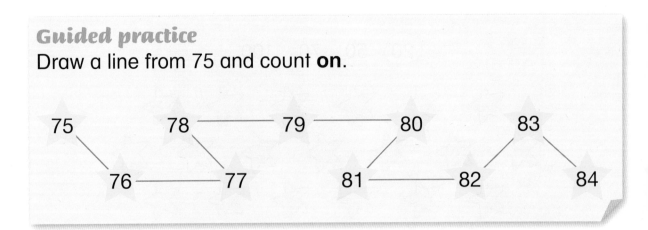

75   78 ——— 79 ——— 80   83

　　76 ——— 77   81 ——— 82   84

Lesson 2: **Reading numbers to 100**

Key words
- **number**
- **digit**
- **numeral**

• Read numbers from 0 to 100

Let's learn

Guided practice

Say the number on each planet. Then draw a line to match each planet and star.

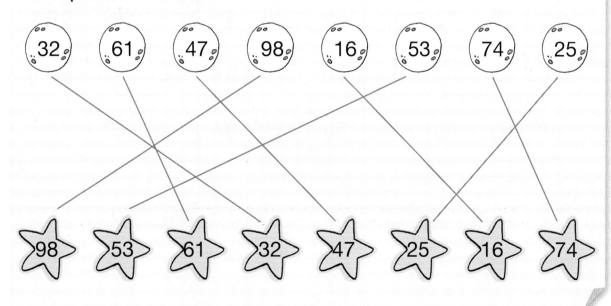

Lesson 3: **Writing numbers to 100**

• Write numbers from 0 to 100

Let's learn

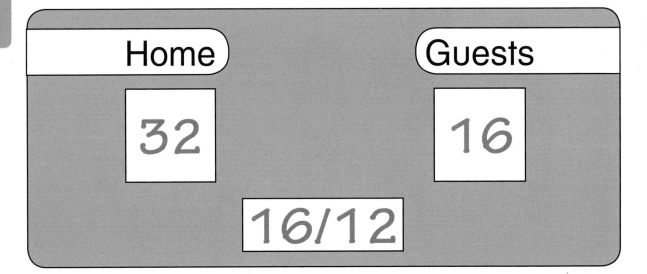

Home		Guests	
32		**16**	
	16/12		

Guided practice

Fill in the missing numbers.

41	42	43	44	45	46	47	48	49	50
51	52	53	54	55	56	57	58	59	60
61	62	63	64	65	66	67	68	69	70

Lesson 4: **Reading and writing number names to 100**

Key words
- **number**
- **digit**
- **number name**
- **numeral**

- Read and write number names from 0 to 100

Let's learn

85

eighty-five

28

twenty-eight

53

fifty-three

60

sixty

Guided practice

Write the matching number name.

| 71 | _seventy-one_ |

Write the matching number.

| forty-eight | _48_ |

Lesson 1: **The link between addition and subtraction**

- Understand that addition is the opposite of subtraction
- Record matching addition and subtraction number sentences

Key words
- addition
- augend
- addend
- sum
- total
- subtraction
- minuend
- subtrahend
- difference
- equals

Let's learn

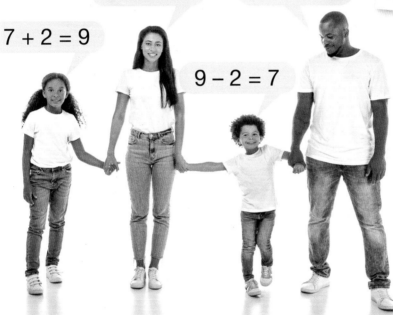

$7 + 2 = 9$

$2 + 7 = 9$

$9 - 7 = 2$

$9 - 2 = 7$

Guided practice

Work out the answer. Then write one addition and two subtractions to match.

$6 + 4 = \boxed{10}$

$\boxed{4} + \boxed{6} = \boxed{10}$

$\boxed{10} - \boxed{4} = \boxed{6}$

$\boxed{10} - \boxed{6} = \boxed{4}$

Lesson 2: **Making 20**

• Know pairs of numbers that total 20

Let's learn

Key words
• **number pairs**
• **number family**
• **add**
• **subtract**
• **total**
• **equals**

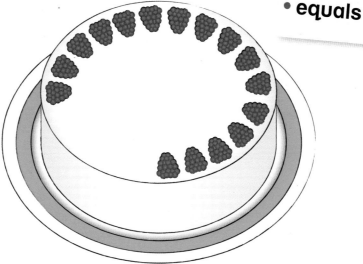

Guided practice

Complete the diagram to show the pair of numbers that total 20. Then complete the number sentences.

20	
16	4

16 + 4 = 20

4 + 16 = 20

20 – 4 = 16

20 – 16 = 4

Lesson 3: **Adding and subtracting tens**

- Add and subtract multiples of 10

Key words
- tens
- multiple of 10
- number pairs
- add
- subtract
- total
- equals

Let's learn

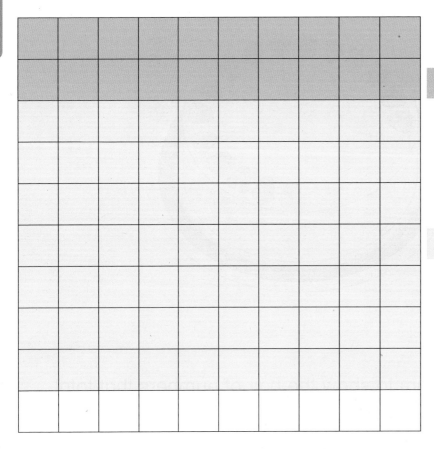

20

70

Guided practice

Write an addition fact and a subtraction fact.

50	
40	10

$$40 + 10 = 50$$

$$50 - 40 = 10$$

Lesson 4: **Adding more than two numbers**

- Add more than two small numbers together

Number

Let's learn

Guided practice

Order the numbers, then add them together.
Estimate the total first.

$1 + 3 + 2 + 1 =$ | 7 | Estimate: | 8 |

| 3 | + | 2 | + | 1 | + | 1 | = | 7 |

Number

Lesson 1: **Adding 2-digit numbers and ones**

- Add a 1-digit number to a 2-digit number

Let's learn

$$45 + 3 =$$

3 ... 4, 5, 6, 7, 8, 9 ...

45 ...
46, 47, 48

$$45 + 3$$

40   5

+ 3

40   8

48

Guided practice

Use the number line to count on from the greater number.
Estimate first.

$4 + 92 =$ 96   Estimate: 95

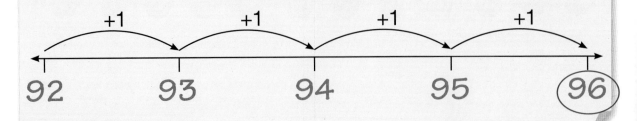

+1   +1   +1   +1

92   93   94   95   96

Lesson 2: **Adding 2-digit numbers and tens**

• Add tens to a 2-digit number

Key words
• tens
• add
• plus
• more
• count on

Number

Let's learn

46 + 30 =

1	2	3	4	5	6	7	8	9	10
11	12	13	14	15	16	17	18	19	20
21	22	23	24	25	26	27	28	29	30
31	32	33	34	35	36	37	38	39	40
41	42	43	44	45	46	47	48	49	50
51	52	53	54	55	56	57	58	59	60
61	62	63	64	65	66	67	68	69	70
71	72	73	74	75	76	77	78	79	80
81	82	83	84	85	86	87	88	89	90
91	92	93	94	95	96	97	98	99	100

46 + 30

40   6

+ 30

70   6

76

Guided practice

Use the number line to count on in tens.

34 + 20 = 54

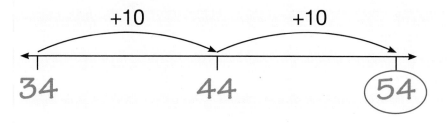

Number

Lesson 3: **Adding 2-digit numbers (1)**

• Add pairs of 2-digit numbers

Let's learn

$$64 + 15 =$$

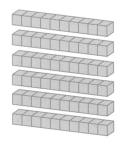   +

Guided practice

Use the number line to add on the ones first, then the tens.

$$52 + 35 = \boxed{87} \qquad \text{Estimate: } \boxed{90}$$

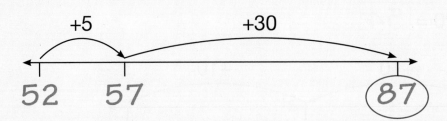

+5   +30

52   57   87

Number

Lesson 4: **Adding 2-digit numbers (2)**

• Add pairs of 2-digit numbers

Key words
• add
• plus
• partition
• tens
• ones
• expanded method
• formal method

Let's learn

$4 \mid 6 \rangle + 2 \mid 3 \rangle =$

$4 \mid 0 \rangle + 6 \rangle + 2 \mid 0 \rangle + 3 \rangle =$

$6 \mid 0 \rangle + 9 \rangle = 69$

Expanded written method

```
  10s 1s
    4   6
+   2   3
        9
    6   0
    6   9
```

Formal written method

```
  10s 1s
    4   6
+   2   3
    6   9
```

Guided practice

Work out the answer using the expanded written method.

Estimate = $\boxed{56}$

```
        3   5
    +   2   1
            6
        5   0
        5   6
```

Lesson 1: **Subtracting 2-digit numbers and ones**

- Subtract a 1-digit number from a 2-digit number

Key words
- subtract
- take away
- count back
- digit
- ones

Let's learn

65 – 2 =

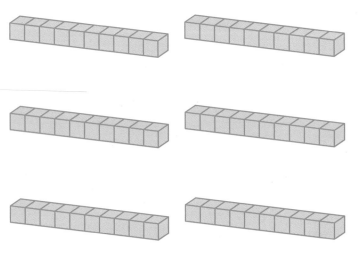

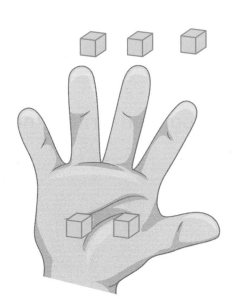

Guided practice

Use the number line to count back. Estimate first.

$87 - 4 =$ 83   Estimate: 82

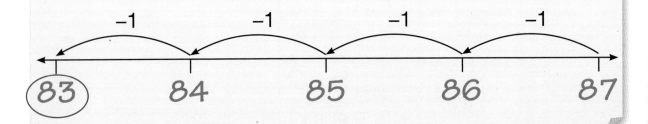

Lesson 2: **Subtracting 2-digit numbers and tens**

• Subtract tens from a 2-digit number

Number

Key words
• subtract
• take away
• count back
• digit
• tens

Let's learn

78 – 40 =

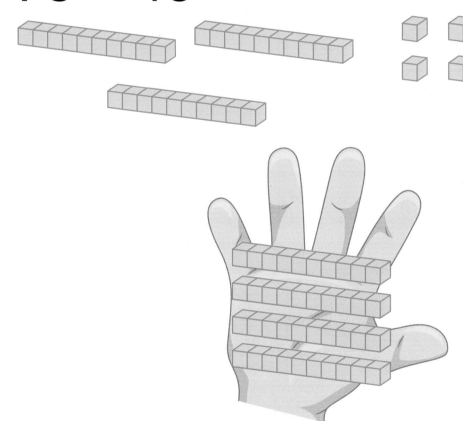

Guided practice

Use the number line to count back in tens.

54 – 20 = $\boxed{34}$

–10   –10

34   44   54

Lesson 3: **Subtracting 2-digit numbers (1)**

• Subtract pairs of 2-digit numbers

Let's learn

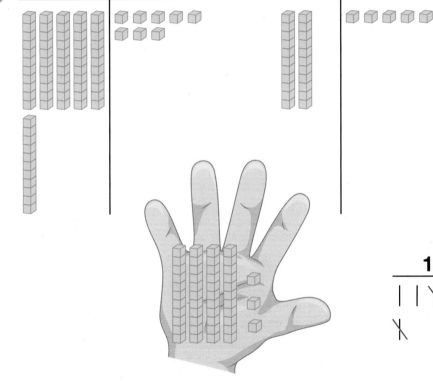

10s	1s

Guided practice

Use the place value chart to solve the subtraction.
Estimate first.

69 – 23 = $\boxed{46}$

Estimate: $\boxed{45}$

Lesson 4: **Subtracting 2-digit numbers (2)**

- Subtract one 2-digit number from another

Let's learn

86 – 34 =

Expanded written method

10s	1s
80	6
– 30	4
50	2

Formal written method

10s	1s
8	6
– 3	4
5	2

Guided practice

Work out the answer using the expanded written method. Estimate first.

75 – 32 = 43

Estimate: 41

10s	1s
70	5
– 30	2
40	3

Lesson 1: **Multiplication as repeated addition (1)**

- Count objects in groups to solve multiplication problems with repeated addition

Key words
- **multiply**
- **multiplied by**
- **times**
- **add**

Let's learn

$$2 + 2 + 2 + 2 = 8$$

$$2 \times 4 = 8$$

Guided practice

Count the fruit in their groups to solve the multiplication.

$2 \times 6 = \boxed{12}$

$= \boxed{12}$

Lesson 2: **Multiplication as repeated addition (2)**

- Use a diagram to solve multiplication problems with repeated addition

Number

Let's learn

$$5 \times 6 =$$

× × × × ×	× × × × ×	× × × × ×	× × × × ×	× × × × ×	× × × × ×

Guided practice

Draw crosses in the diagram to solve the multiplication.

$2 \times 4 =$ 8

Lesson 3: **Multiplication using a number line**

Key words
- **multiply**
- **multiplied by**
- **times**
- **add**

- Use a number line to solve multiplication problems with repeated addition

Let's learn

$2 \times 9 =$

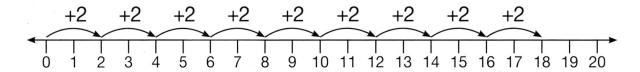

Guided practice

Draw the correct number of jumps to match the calculation. Then write the answer.

$2 \times 8 = \boxed{16}$

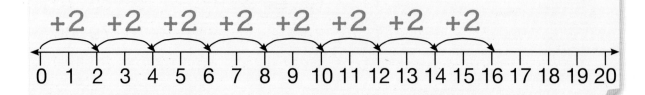

Lesson 4: **Multiplication using a 100 square**

Key words
- multiply
- multiplied by
- times
- add

- Use a 100 square to solve multiplication problems with repeated addition

Let's learn

$5 \times 9 =$

1	2	3	4	5	6	7	8	9	10
11	12	13	14	15	16	17	18	19	20
21	22	23	24	25	26	27	28	29	30
31	32	33	34	35	36	37	38	39	40
41	42	43	44	45	46	47	48	49	50
51	52	53	54	55	56	57	58	59	60
61	62	63	64	65	66	67	68	69	70
71	72	73	74	75	76	77	78	79	80
81	82	83	84	85	86	87	88	89	90
91	92	93	94	95	96	97	98	99	100

Guided practice

Use the 100 square to help you to solve the multiplication.

$5 \times 8 = \boxed{40}$

1	2	3	4	5	6	7	8	9	10
11	12	13	14	15	16	17	18	19	20
21	22	23	24	25	26	27	28	29	30
31	32	33	34	35	36	37	38	39	40
41	42	43	44	45	46	47	48	49	50
51	52	53	54	55	56	57	58	59	60
61	62	63	64	65	66	67	68	69	70
71	72	73	74	75	76	77	78	79	80
81	82	83	84	85	86	87	88	89	90
91	92	93	94	95	96	97	98	99	100

Lesson 1: **Multiplication as an array (1)**

- Understand multiplication as an array

Let's learn

$$5 \times 3 = 15$$
$$3 \times 5 = 15$$

Guided practice

Write and solve a number sentence for the array.

$$\boxed{3} \times \boxed{4} = \boxed{12}$$

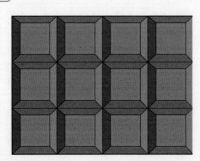

Lesson 2: **Multiplication as an array (2)**

Key words
• **multiply**
• **multiplied by**
• **times**
• **array**

Number

• Draw arrays to solve multiplication problems

Let's learn

$$4 \times 2 =$$

Guided practice

Draw an array to match the number sentence.

$2 \times 7 =$ 14

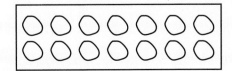

Number

Lesson 3: **The equals sign**

- Understand that an array can show two multiplications with the same answer
- Understand that the facts on either side of the equals sign have the same value

Let's learn

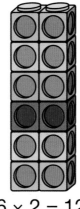

$6 \times 2 = 12$ $2 \times 6 = 12$

$$6 \times 2 = 2 \times 6$$

Guided practice

Write an equal number statement to match the array.

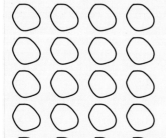 $\boxed{6} \times \boxed{4} = \boxed{4} \times \boxed{6}$

Lesson 4: **Solving problems (1)**

Key words
- **multiply**
- **multiplied by**
- **times**
- **array**

Number

- Use arrays to solve 'real-life' multiplication problems

Let's learn

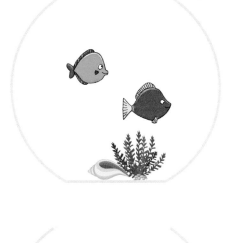

Guided practice

Solve the multiplication problem. Draw an array in the box to help you.

3 friends pick 2 strawberries each.

How many strawberries do they pick altogether? | 6 |

Lesson 1: **Division – sharing between 2**

• Share up to 20 objects between 2

Let's learn

Key words
• divide
• divide by
• divide between
• share equally
• share between

Guided practice

Share the watermelon between 2 plates.

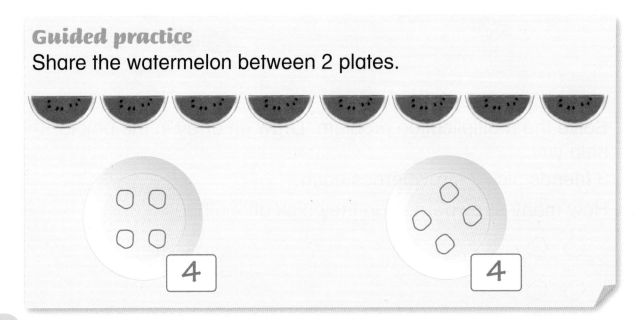

4   4

Lesson 2: **Division – sharing between more than 2**

• Share amounts between more than 2

Let's learn

Key words
• divide
• divide by
• divide between
• share equally
• share between

15 marbles

Guided practice

Share 30 apples equally between 5 baskets.

How many apples are in each basket? $\boxed{6}$

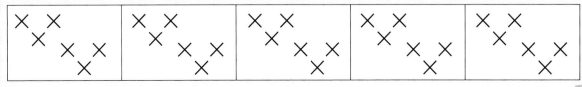

Lesson 3: **The division sign (1)**

- Recognise the division sign
- Share to solve division number sentences

Let's learn

$8 \div 2 = 4$

Guided practice

Solve the division. Use the sharing diagram to help you.

$30 \div 5 = \boxed{}$

X X X X X X X X X X X X X X X X X X X X X X X X X X X X X X				
X X X X X X	X X X X X X	X X X X X X	X X X X X X	X X X X X X

Lesson 4: **Solving problems (2)**

- Solve 'real-life' division problems by sharing

Key words
- divide
- divide by
- divide between
- share equally
- share between

Number

Let's learn

Divide the chairs equally between the tables. How many chairs are at each table?

Guided practice

Use counters or a sharing diagram to solve the problem.

There are 15 children and 5 paddling pools. Share the children equally. How many children are in each paddling pool?

$$15 \div 5 = 3$$

X	X	X	X	X	X	X	X	X	X	X	X	X	X	X
X X X	X X X	X X X	X X X	X X X

Number

Lesson 1: **Division as grouping (1)**

Key words
* **divide**
* **divide by**
* **divide into**
* **groups**

* Group objects to discover how many groups of 2 there are

Let's learn

Guided practice

Draw a ring around each group of 2 to find out how many groups there are.

How many groups of 2 in 10?　5

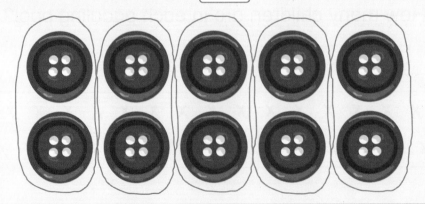

Lesson 2: **Division as grouping (2)**

- Discover how many groups of 5 or 10 there are

Key words
- **divide**
- **divide by**
- **divide into**
- **groups**

Let's learn

Guided practice

Draw groups of 10 dots to find how many groups of 10 are in 20.

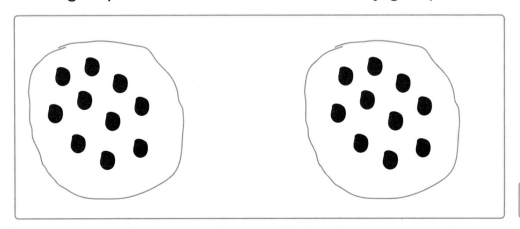

2

Lesson 3: **The division sign (2)**

- Recognise the division sign
- Solve division number sentences by grouping

Key words
- **divide**
- **divide by**
- **divide into**
- **groups**

Let's learn

$$15 \div 5 = 3$$

15 biscuits

Guided practice

Solve the division. Draw groups to help you.

$35 \div 5 = \boxed{7}$

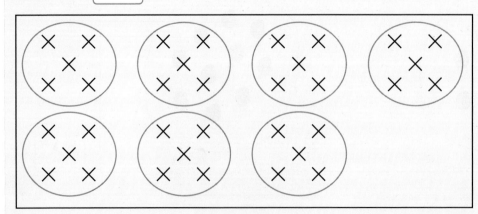

Lesson 4: **Solving problems (3)**

Key words
- **divide**
- **divide by**
- **divide into**
- **groups**

- Solve 'real-life' problems by grouping

Number

Let's learn

Lia has 30 seeds. 5 seeds are planted in each pot. How many pots does Lia need?

Guided practice

Solve the problem using counters or by drawing groups.

20 children go on a school trip. Two children fit on each seat on the bus. How many seats do they need?

$$20 \div 2 = 10$$

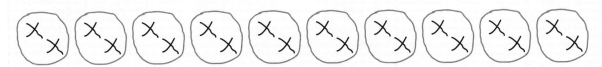

Lesson 1: **Division as repeated subtraction**

• Use objects to solve division problems with repeated subtraction

Let's learn

$$6 \div 2 =$$

Guided practice

Subtract groups of 2 sharks by crossing them out until you have no sharks left. Then count how many groups of sharks you subtracted.

$16 \div 2 =$ $\boxed{8}$

Lesson 2: **Division using a number line (1)**

> **Key words**
> • **divide**
> • **subtract**
> • **number line**
> • **jumps**

- Use a number line to solve division problems with repeated subtraction

Let's learn

$20 \div 5 =$

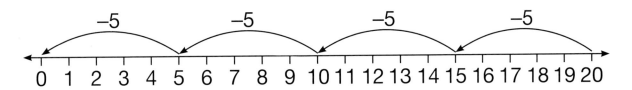

Guided practice

Subtract groups on the number line. Count the jumps to find the answer.

$70 \div 10 = \boxed{7}$

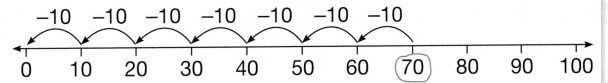

$16 \div 2 = \boxed{8}$

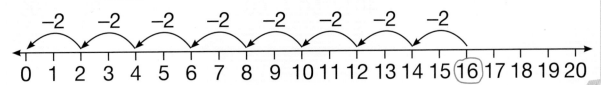

Lesson 3: **Division using a number line (2)**

- Use a number line to solve division problems with repeated subtraction

Key words
- **divide**
- **divided by**
- **subtract**
- **equals**
- **number line**
- **jumps**

Let's learn

$$14 \div 2 = 7$$

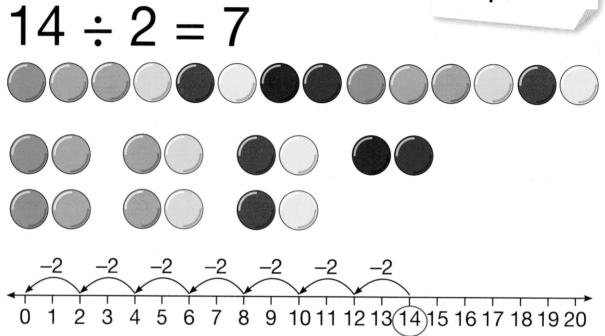

Guided practice

Use the number line to solve the division.

$$40 \div 10 = \boxed{4}$$

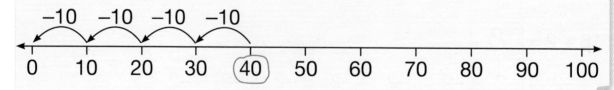

Lesson 4: **Solving problems (4)**

Key words
- divide
- problem
- subtract
- equals
- number line
- jumps

- Solve 'real-life' division problems using repeated subtraction

Let's learn

There are 15 balloons. How many children can have 5 balloons each?

Guided practice

Draw a number line to subtract groups to solve the problem.

There are 8 sleepy kittens. 2 kittens can fit in each cat bed.

How many cat beds do they need? 4

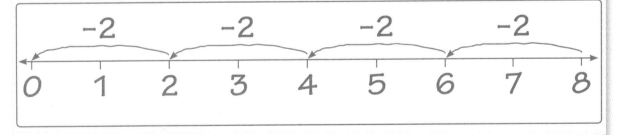

Lesson 1: **2 times table: multiplication facts**

- Recall multiplication facts for the 2 times table

Let's learn

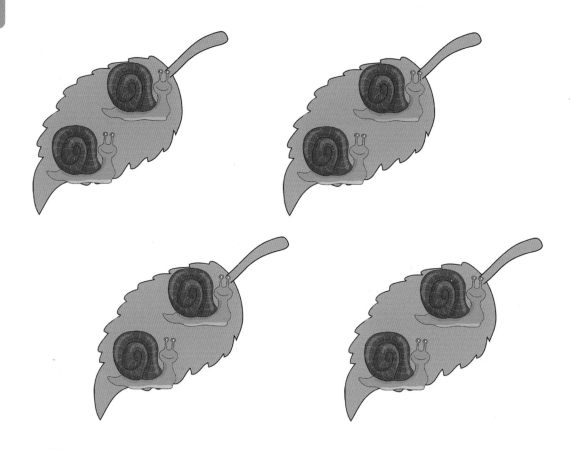

Guided practice

Use the number line to answer the 2 times table fact.

$2 \times 9 = \boxed{18}$

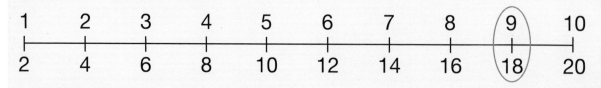

Lesson 2: **2 times table: division facts**

• Recall division facts for the 2 times table

Key words
• divide
• multiply
• times
• twos

Number

Let's learn

Guided practice

Use the number line to answer the 2 times table division fact.

$8 \div 2 =$ | 4 |

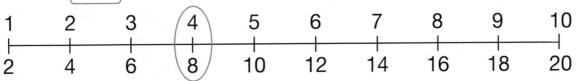

Number

Lesson 3: **5 times table: multiplication facts**

- Recall multiplication facts for the 5 times table

Let's learn

Guided practice

Use the number line to answer the 5 times table fact.

$5 \times 8 = \boxed{40}$

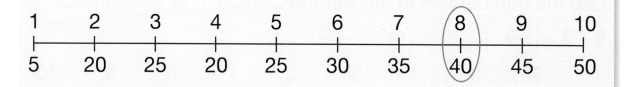

1	2	3	4	5	6	7	8	9	10
5	20	25	20	25	30	35	40	45	50

Lesson 4: **5 times table: division facts**

• Recall division facts for the 5 times table

Number

Let's learn

  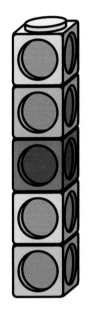   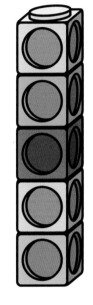

Guided practice

Use the number line to answer the 5 times table division fact.

10 ÷ 5 = [2]

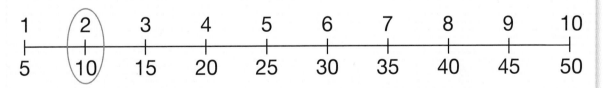

Lesson 1: **10 times table: multiplication facts**

- Recall multiplication facts for the 10 times table

Let's learn

Guided practice

Use the number line to answer the 10 times table fact.

$10 \times 2 =$ ⬚ 20

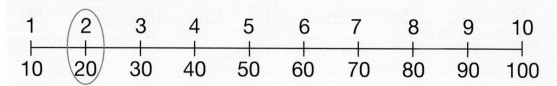

Lesson 2: **10 times table: division facts**

- Recall division facts for the 10 times table

Number

Let's learn

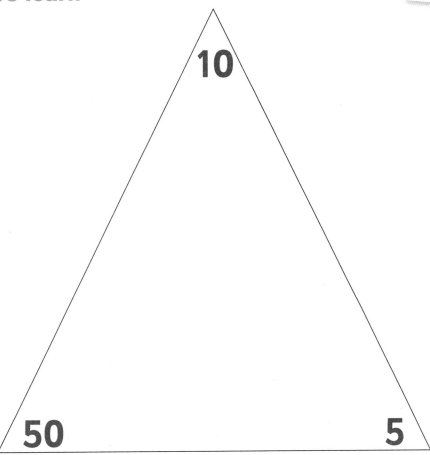

Guided practice

Use the number line to answer the 10 times table division fact.

$60 \div 10 =$ 6

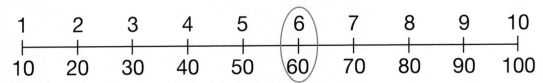

Lesson 3: **1 times table: multiplication and division facts**

Key words
* ones
* multiply
* times
* divide

* Recall multiplication and division facts for the 1 times table

Let's learn

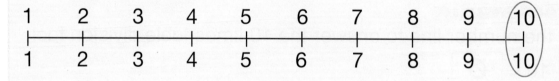

Guided practice

Use the number line to answer the 1 times table multiplication and division facts.

$1 \times 7 = \boxed{7}$

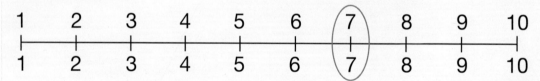

$10 \div 1 = \boxed{10}$

Lesson 4: **1, 2, 5 and 10 times tables**

Key words
- divide
- multiply
- times
- twos
- fives
- tens

- Recall multiplication and division facts for the 1, 2, 5 and 10 times tables

Let's learn

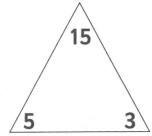

Guided practice

Use the times table grid to answer the facts.

×	1	2	3	4	5	6	7	8	9	10
1	1	2	3	4	5	6	7	8	9	10
2	2	4	6	8	10	12	14	16	18	20
5	5	10	15	20	25	30	35	40	45	50
10	10	20	30	40	50	60	70	80	90	100

a $10 \times 6 = \boxed{60}$ **b** $2 \times 9 = \boxed{18}$ **c** $30 \div 5 = \boxed{6}$

Number

Lesson 1: **Recognising local currency**

- Recognise the currency symbol in local currency
- Recognise the value of coins and notes in local currency

Let's learn

Guided practice

Draw two coins from your local currency. Make sure that you include the currency (if it's shown) and the value of each coin.

Lesson 2: **Paying with dollars and cents**

Key words
- cent
- dollar
- coins
- notes
- value

- Recognise all dollar and cent coins and notes
- Match values of coins and notes to prices

Let's learn

 1 5 10 25 50

Guided practice

Draw a line to match each coin and note to the item with the same value.

 5

 25

 5c

$5

 25c

Number

Lesson 3: **Comparing values**

- Compare and order values of coins and notes

Let's learn

My ice cream costs $1.

My ice cream costs $5.

Guided practice

Colour the coin with the highest value and draw a ring around the coin with the lowest value.

 25   5 $1

Lesson 4: **Equal values**

• Find coins or notes of the same value

Number

Key words
• **same**
• **value**
• **total**

Let's learn

Guided practice

Draw a line to match the total in the purse with the note.

Lesson 1: **Tens and ones**

- Know how many tens and ones are in a 2-digit number

Key words
- **tens**
- **ones**

Let's learn

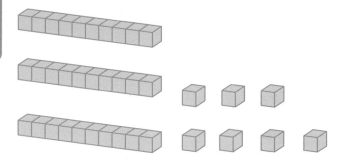

37

45

Guided practice

How many tens? How many ones?

62

tens: ones:

| 6 | 2 |

Lesson 2: **Partitioning**

- Partition 2-digit numbers into tens and ones

🔍 **Key words**
- partition
- tens
- ones

Let's learn

23 is made up to 2 tens and 3 ones.

| **2** | **3** |

| **2** | **0** | **3**

58 is made up to 5 tens and 8 ones.

| **5** | **8** |

| **5** | **0** | **8**

Guided practice

Partition this number.

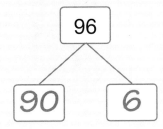

96
90   6

Lesson 3: **Comparing numbers**

- Compare 2-digit numbers

Let's learn

Guided practice

Colour the greatest number.

67   39   71

Lesson 4: **Ordinal numbers**

Key word
• ordinal
  numbers

• Use ordinal numbers

Let's learn

M	T	W	T	F	S	S
April						
1st	2nd	3rd	4th doctor's	5th	6th	7th
8th	9th	10th party	11th	12th	13th	14th
15th	16th	17th	18th	19th	20th	21st
22nd	23rd	24th	25th	26th	27th	28th
29th	30th					

Guided practice

Colour the 8th shell blue.

Number

65

Lesson 1: **Composing and decomposing numbers**

Key words
• tens
• ones
• partition
• number sentence

• Compose and decompose numbers using tens and ones

Let's learn

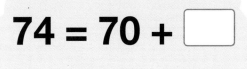

$$74 = 70 + \boxed{}$$

$$58 = \boxed{} + 8$$

Guided practice

Write a number sentence to match the number.

$$\boxed{50} + \boxed{7} = 57$$

Lesson 2: **Regrouping numbers**

Key words
- **tens**
- **ones**
- **number sentence**
- **regroup**

- Regroup 2-digit numbers

Let's learn

$$30 + 6$$

$$30 + 2 + 2 + 2$$

$$10 + 10 + 10 + 6$$

$$30 + 5 + 1$$

Guided practice

Complete each number sentence.

$58 = 50 + \boxed{8}$

$58 = 50 + \boxed{4} + \boxed{4}$

Write a different way to show 58.

$$\boxed{10 + 10 + 10 + 10 + 10 + 8 = 58}$$

Number

Number

Lesson 3: **Ordering 2-digit numbers**

Key words
- order
- smallest
- greatest

• Compare and order 2-digit numbers

Let's learn

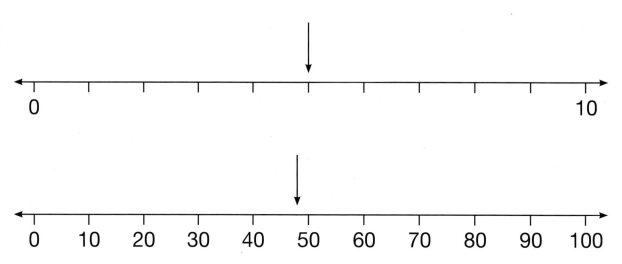

Guided practice

Order the numbers, starting with the **smallest**.

29 46 82 16 55

| 16 | 29 | 46 | 55 | 82 |

Write the number on the number line.

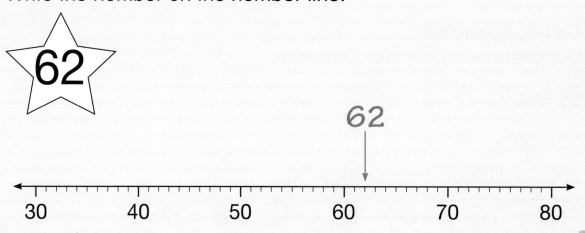

62

62

Number

Lesson 4: **Rounding 2-digit numbers**

- Round 2-digit numbers to the nearest 10

Let's learn

How many people will be at lunch?

About 8.

I'll buy enough food for 10 people, then.

Guided practice

Round the number to the nearest 10.

82   80

Lesson 1: **Quarters of shapes**

Key words
• **quarter**
• **equal**

- Recognise which shapes are divided into quarters and which shapes are not

Let's learn

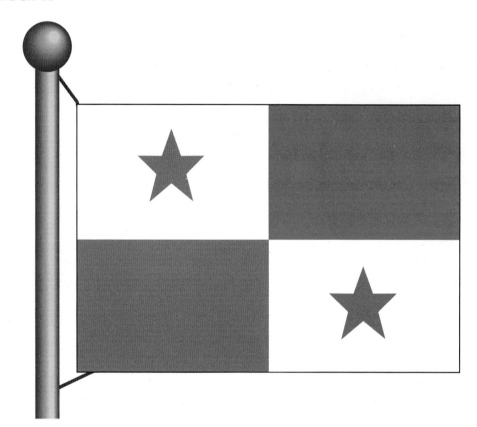

Guided practice

Colour the shapes that are divided into quarters. Then label each quarter $\frac{1}{4}$.

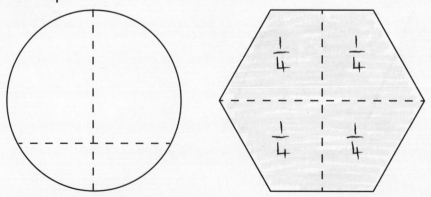

Number

Lesson 2: **Finding one quarter**

Key words
- **quarter**
- **equal**
- **whole**

• Find one quarter of a shape

Let's learn

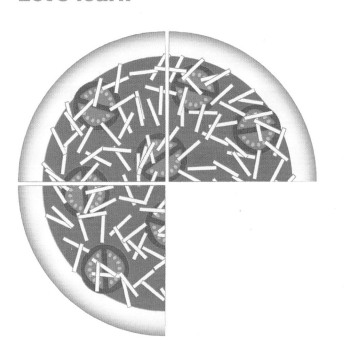

Guided practice

Draw lines to divide the shape into quarters. Then colour $\frac{1}{4}$ of the shape.

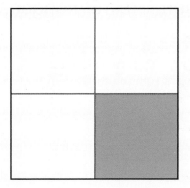

Lesson 3: **Quarters of sets of objects**

• Find one quarter of a set of objects

Let's learn

Key words
• quarter
• whole
• share
• divide
• equal

Guided practice

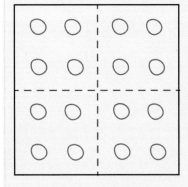

Draw dots in each quarter to find $\frac{1}{4}$.
There are 16 mangoes in a bag. Olivia takes $\frac{1}{4}$ of them.
How many mangoes does Olivia take?

4

Lesson 4: **Finding one quarter of a set of objects**

Key words
* quarter
* whole
* share
* divide
* multiply
* equal

Number

* Use one quarter of a set of objects to work out how many objects are in the whole set

Let's learn

$$\frac{1}{4}$$

Guided practice

Draw the rest of the fruit. Then count how many there are altogether to find how many are in the full set.

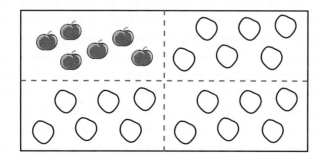

24

Lesson 1: **Equal fractions**

- Recognise how big fractions are, compared to each other
- Recognise fractions that equal the same amount

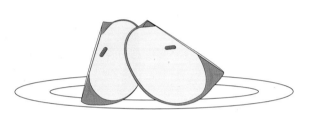

Key words
- whole
- half
- halves
- quarters
- equal

Let's learn

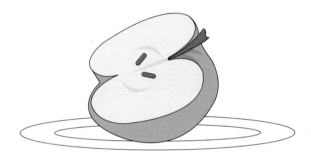

Guided practice

Draw rings around the fractions that match the circle.

$\frac{1}{2}$　1　$\frac{2}{2}$　$\frac{2}{4}$　$\frac{1}{4}$　$\frac{4}{4}$

Lesson 2: **Combining fractions**

- Combine halves and quarters to create new fractions

Key words
- half
- halves
- quarters
- three-quarters
- whole
- equal

Let's learn

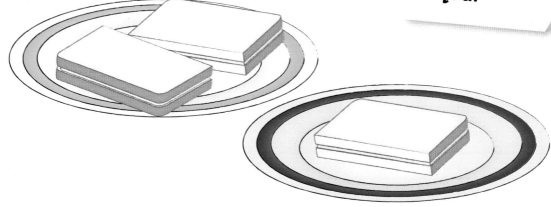

Guided practice

Draw a ring around the fraction that matches the diagram.

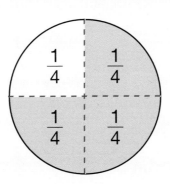

1      $\frac{3}{4}$ $\frac{3}{2}$

Number

Lesson 3: **Fractions as operators**

- Find one half or one quarter of a shape or an amount to 20

Let's learn

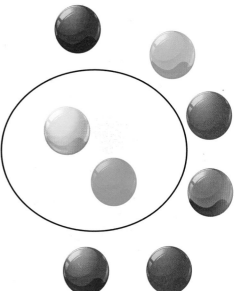

Guided practice

Colour $\frac{1}{4}$ of the stars.

Lesson 4: **Fractions as division**

Key words
- fraction
- half
- quarter
- divided by
- equals

Number

- Understand fractions as division

Let's learn

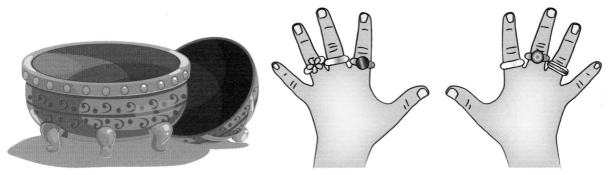

Guided practice

Complete the number sentences to solve the problem.

There are 14 skipping ropes in the playground. You tidy half of them away. How many skipping ropes are left?

$\boxed{14} \div \boxed{2} = \boxed{7}$

Lesson 1: **Using a calendar**

- Use and interpret a calendar

Key words
- calendar
- months
- days
- years

Let's learn

Geometry and Measure

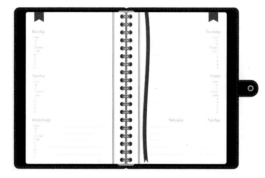

2022

Guided practice

Use the calendar for June to answer these questions.

June						
Monday	Tuesday	Wednesday	Thursday	Friday	Saturday	Sunday
1 *football practice*	2	3	4	5 *gymnastics*	6	7
8 *football practice*	9	10	11	12	13	14

What date is gymnastics? <u>5th</u>

What day of the week is football practice? <u>Mondays</u>

Lesson 2: **Ordering time**

• Order units of time

Let's learn

Key words
• **years**
• **months**
• **weeks**
• **days**
• **hours**
• **minutes**
• **seconds**

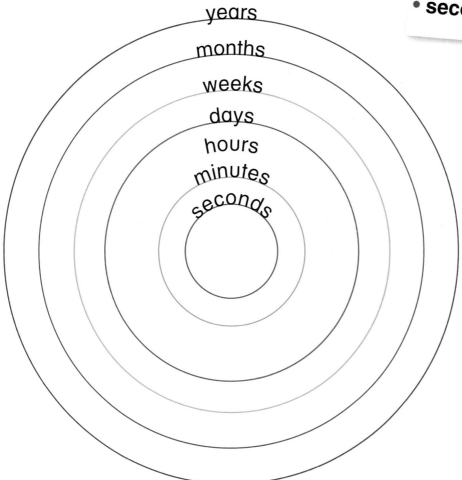

Geometry and Measure

Guided practice

Number these units of time from shortest (1) to longest (3).

hour 1   week 3   day 2

Lesson 3: **Reading and showing the time (analogue)**

• Read and record the time to 5 minutes on an analogue clock

Key words
• time
• clock
• analogue
• hours
• minutes

Let's learn

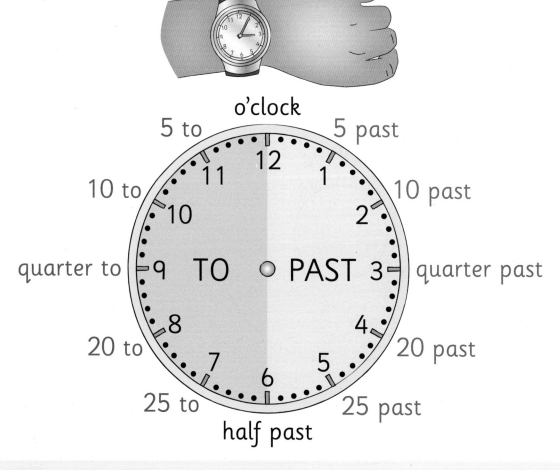

o'clock
5 to   5 past
10 to   10 past
quarter to   TO   PAST   quarter past
20 to   20 past
25 to   25 past
half past

Guided practice

Write the time.

_10 to 5_

80

Lesson 4: **Reading and showing the time (digital)**

* Read and record the time to 5 minutes on a digital clock

Key words
* **time**
* **clock**
* **digital**
* **hours**
* **minutes**

Let's learn

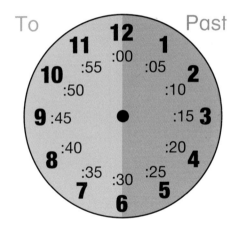

Guided practice
Write the digital time to match.

4:10

Lesson 1: **2D shapes (1)**

• Identify, describe and sort 2D shapes

Let's learn

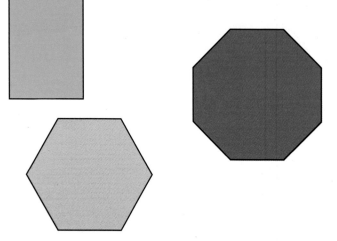

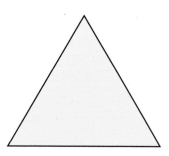

Guided practice

Sort the shapes by writing the letters in the correct box.

6 or more sides	fewer than 6 sides
A G	B C D E F

Lesson 2: **2D shapes (2)**

Key words
- **square**
- **circle**
- **rectangle**
- **triangle**
- **pentagon**
- **hexagon**
- **octagon**
- **tessellate**
- **pattern**

- Identify 2D shapes in familiar objects
- Explore patterns using shapes

Let's learn

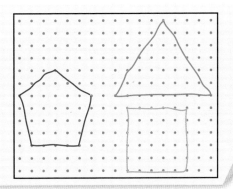

Guided practice

Use the geoboard to make a triangle, a square and a pentagon.

Then copy them here.

Lesson 3: **Lines of symmetry**

- Recognise horizontal and vertical lines of symmetry

Geometry and Measure

Let's learn

Guided practice

Colour the shape that has both vertical and horizontal lines of symmetry.

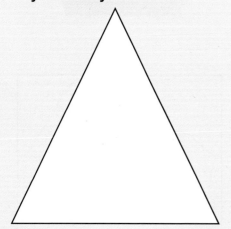

Lesson 4: **Angles**

- Identify right angles
- Understand that a right angle is a quarter turn

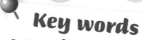

Key words
- angle
- right angle
- turn
- quarter
- full
- whole

Geometry and Measure

Let's learn

Guided practice

Make a list of things you can see in the classroom that have right angles.

- <u>book</u>
- <u>picture frame</u>

85

Lesson 1: **Recognising 3D shapes**

- Identify and name 3D shapes
- Recognise 3D shapes in familiar objects

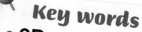

Key words
- **3D**
- **shape**
- **cube**
- **cuboid**
- **cylinder**
- **pyramid**
- **sphere**

Let's learn

Guided practice

Match the object to its shape.

86

Lesson 2: **Describing 3D shapes**

Key words
- **3D**
- **shape**
- **face**
- **edge**
- **corners**
- **vertex**
- **vertices**

- Talk about the faces, edges and vertices of 3D shapes

Let's learn

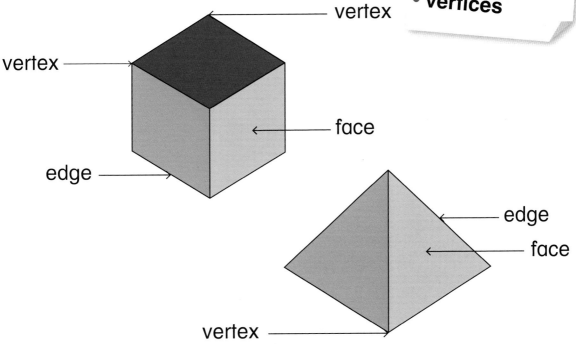

Geometry and Measure

Guided practice

Complete the table.

Name of shape	Number of faces	Number of vertices	Number of edges
cylinder	3	0	2

Lesson 3: **Sorting 3D shapes**

- Compare and sort 3D shapes

Key words
- face
- edge
- vertex
- vertices
- flat
- curved
- same
- different

Geometry and Measure

Let's learn

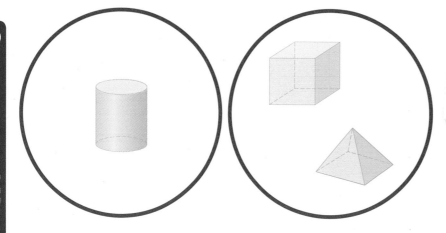

Guided practice

Colour red all the shapes with one or more curved faces.
Colour blue all the shapes with only flat faces.

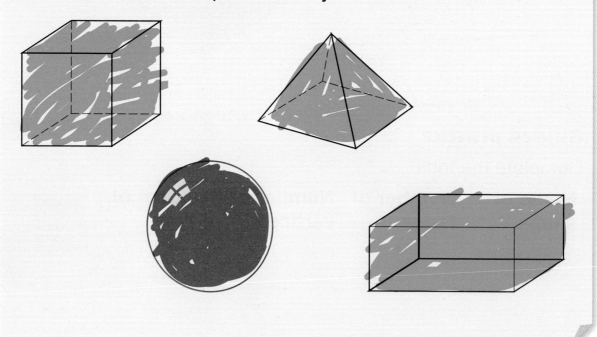

Lesson 4: **Making 3D shapes**

• Make 3D shapes

Key words
• face
• edge
• vertex
• vertices
• model
• base

Geometry and Measure

Let's learn

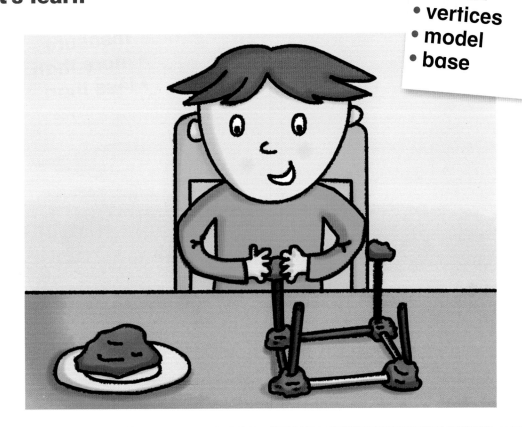

Guided practice

Use straws and modelling clay to make a pyramid.

Draw your pyramid.

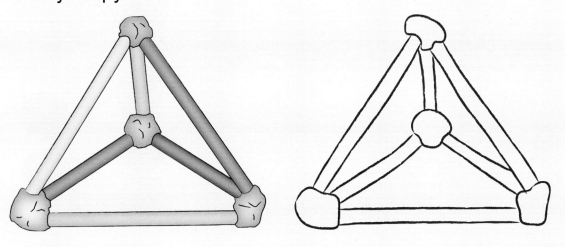

Lesson 1: **Measuring length with non-standard units**

- Estimate and measure length with units of measure that are the same

Key words
- length
- estimate
- measure
- unit of measure
- more than
- less than

Let's learn

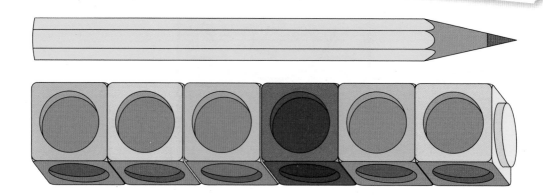

Guided practice

Estimate the length of the pencil in counters.
Then use counters to measure the pencil.

Estimate:

| 6 |

Length:

| 7 |

Lesson 2: **Centimetres and metres**

- Recognise and use the standard units: centimetres and metres

Key words
- metre
- centimetre
- ruler
- metre rule

Geometry and Measure

Let's learn

Guided practice

Draw one example in each box.

Shorter than 30 centimetres	Taller than 30 centimetres	Taller than 1 metre

Lesson 3: **Drawing and measuring lines**

• Draw and measure lines with a ruler

Let's learn

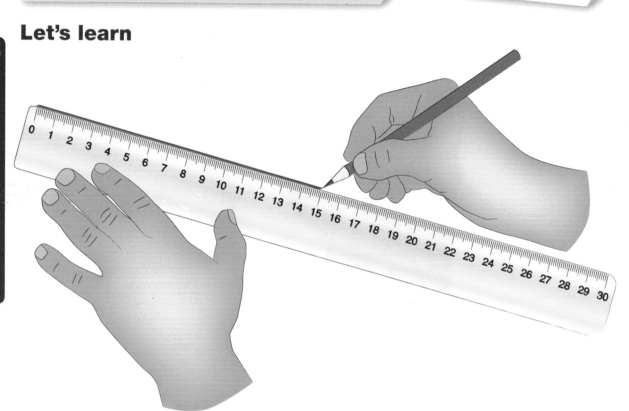

Guided practice

Draw a line 9 centimetres long.

_____

Lesson 4: **Estimating length**

- Estimate lengths of objects in centimetres

Key words
- **estimate**
- **centimetre**

Let's learn

Guided practice

Estimate the height of the candy stick, then measure it.

Estimate:

| 6 |

Height:

| 5 |

Lesson 1: **Measuring mass with non-standard units**

- Estimate and measure mass with units of measure that are the same

Key words
- mass
- estimate
- unit of measure
- level
- balance scale

Let's learn

Guided practice

Find the mass of an object. Use the balance scale to show what you did. Estimate the mass of the object first.

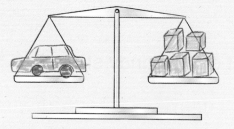

Estimate: | 6 | cubes

Mass: | 5 | cubes

Lesson 2: **Grams and kilograms**

Key words
- **gram**
- **kilogram**

- Recognise and use the standard units: grams and kilograms

Let's learn

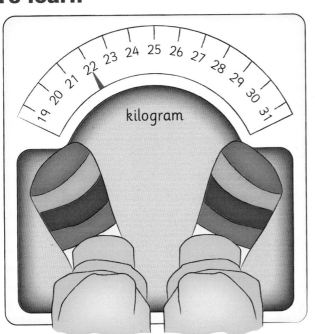

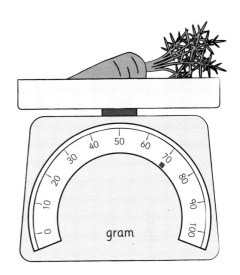

Guided practice

Draw a ring around the measurement you would use for each object.

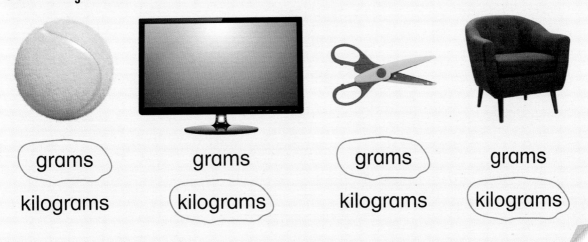

grams	grams	grams	grams
kilograms	kilograms	kilograms	kilograms

Lesson 3: **Measuring mass with grams and kilograms**

Key words
- grams
- kilograms
- measure

- Measure mass in grams and kilograms

Let's learn

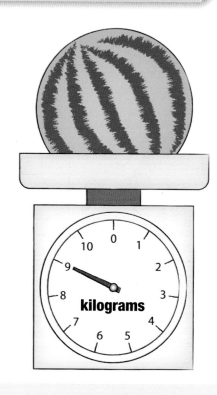

Guided practice

What is the mass of each bag of food?

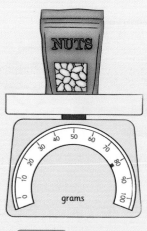

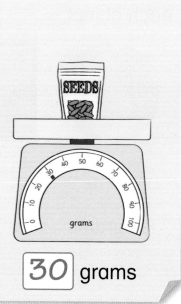

| 3 | kilograms | 80 | grams | 30 | grams |

Lesson 4: **Comparing mass**

• Compare mass in grams and kilograms

Key words
• **compare**
• **kilograms**
• **grams**

Let's learn

Geometry and Measure

Guided practice

Write the mass of each box of food. Then complete the sentence.

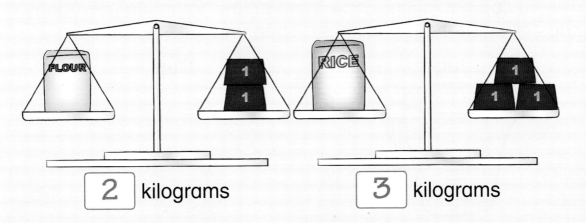

| 2 | kilograms |

| 3 | kilograms |

The rice is 1 kilogram heavier than the flour.

Lesson 1: **Measuring capacity with non-standard units**

Key words
- capacity
- measure
- unit of measure

- Estimate and measure capacity with units of measure that are the same

Geometry and Measure

Let's learn

Guided practice

Take a cup and a jug. Draw them below. Estimate how many cups of water the jug will hold. Then find out.

Jug	Cup

Estimate: 10 cups   Capacity: 13 cups

98

Lesson 2: **Litres and millilitres**

Key words
- litre
- millilitre

- Recognise and use the standard units: litres and millilitres

Let's learn

Guided practice

Colour each container to show how much water is in it.

4 litres

5
4
3
2
1

litre

50 millilitres

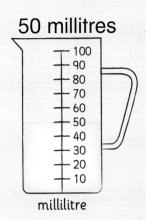

100
90
80
70
60
50
40
30
20
10

millilitre

Lesson 3: **Measuring capacity using litres and millilitres**

- Estimate and measure capacity in litres and millilitres

Let's learn

Guided practice

Look at your containers. Draw them below to show how much water each one holds.

Holds **less than** half a litre	Holds about half a litre	Holds **more than** half a litre

Lesson 4: **Measuring temperature**

- Compare temperatures
- Read and interpret the scale on a thermometer

Key words
- compare
- thermometer
- temperature
- degrees
- hot
- cold
- warm
- cool
- high
- low

Let's learn

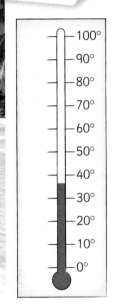

Guided practice

Write the temperature to the nearest 10 degrees.

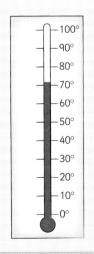

[70] degrees

Lesson 1: **Drawing reflections**

Key words
• reflection
• same
• shape

• Sketch the reflection of a 2D shape

Let's learn

Guided practice

Sketch the reflection of the shape. Use a mirror to help you.

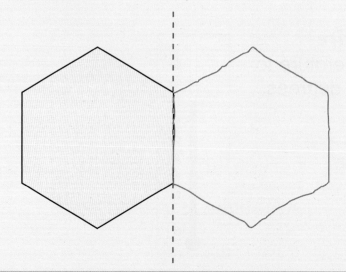

Geometry and Measure

Lesson 2: **Position**

- Give and follow instructions for position

Let's learn

The 🐐 is behind the 🌳

Key words
- **in front of**
- **next to**
- **top**
- **middle**
- **below**
- **behind**
- **above**
- **underneath**
- **between**
- **on**

Geometry and Measure

Guided practice

on   in front of   behind

The house is __behind__ the man.

The tree is __in front of__ the house.

The man is sitting __on__ the wall.

Lesson 3: **Direction and movement**

• Follow and give directions to move from one position to another

Key words
• **left**
• **right**
• **through**
• **straight**
• **forward**
• **go back**

Let's learn

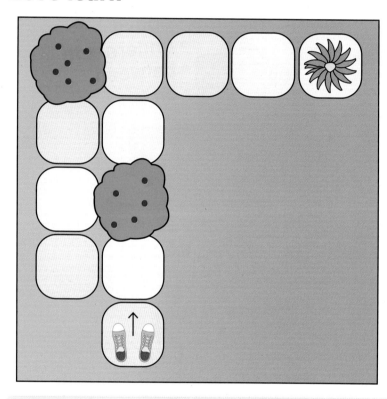

Guided practice

Help the beetle get to the hole by drawing a line. Then write the directions. Use the code below.

Forward = F   Right = R   Left = L

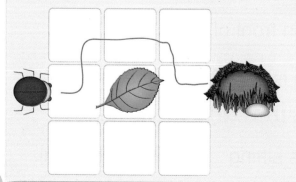

F 1
Turn L, then F 1
Turn R, then F 2
Turn R, then F 1
Turn L, then F 1

Lesson 4: **Whole, half and quarter turns**

• Make whole, half and quarter turns clockwise and anticlockwise

 Key words
• whole turn
• half turn
• quarter turn
• clockwise
• anticlockwise

Geometry and Measure

Let's learn

 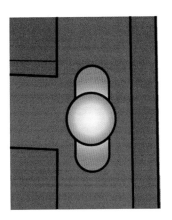 

Guided practice
Use a red pencil to mark the direction on the circle.
a quarter turn clockwise

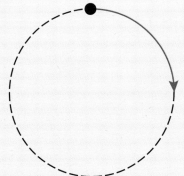

105

Lesson 1: **Using tally charts to collect data**

Key words
- collect
- survey
- data
- tally chart
- tally

- Use a tally chart to collect and interpret data

Let's learn

Guided practice

Use the tally chart to answer the questions.

Fruit	Tally
banana	卌 I
apple	I I I I
orange	卌 I I I I
grapes	卌 I

How many?

a oranges $\boxed{9}$

b apples $\boxed{4}$

c bananas $\boxed{6}$

Statistics and Probability

Lesson 2: **Pictograms and block graphs**

• Use a pictogram or block graph to present data

Let's learn

Vehicles passing our school

Vehicle	Number
car	🚗🚗🚗🚗🚗🚗🚗🚗🚗🚗
truck	🚚🚚🚚🚚
bus	🚌🚌🚌🚌🚌🚌🚌
motorbike	🏍️🏍️🏍️🏍️🏍️
bicycle	🚲🚲🚲

Vehicles passing our school

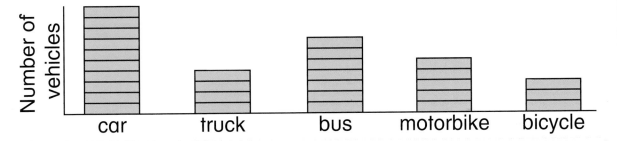

Guided practice

Draw a pictogram for the weather this week.

Weather this week

Weather	Number of days
rain	2
sun	3
cloudy	1
windy	1

Weather this week

Weather	Number of days
rain	🌧️🌧️
sun	☀️☀️☀️
cloudy	☁️
windy	🌬️

Statistics and Probability

Lesson 3: **Venn diagrams**

- Read and create Venn diagrams with two sorting rules

Key words
- Venn diagram
- sort

Let's learn

Animals and legs

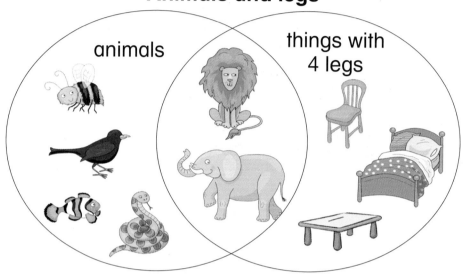

Guided practice

Work out the two sorting rules, then write the labels.

Birds and flying

Does the Venn diagram tell us all the things in the world that fly?

yes (no)

Why/why not? It only shows us the data that was added. Not all things that fly were in the data.

Lesson 4: **Carroll diagrams**

• Read and create Carroll diagrams with two sorting rules

Key words
• **Carroll diagram**
• **sort**

Let's learn

Birds that can fly

	is a bird	is not a bird
can fly	eagle	bat
cannot fly	penguin	cat

Statistics and Probability

Guided practice

Draw at least one piece of food in each part of the Carroll diagram.

Things we eat

	is red	is not red
is a fruit	(berries)	banana
is not a fruit	jelly	sandwich

109

Lesson 1: **Identifying patterns**

- Recognise and describe regular and random patterns

Key words
- **pattern**
- **regular**
- **random**
- **next**

Let's learn

Guided practice

Look at these four patterns. Cross out the random pattern.
Continue the other three patterns.

a

b

c

d

Statistics and Probability

110

Lesson 2: **Chance**

- Look at patterns in chance experiments

Let's learn

Guided practice

Colour the stars so that you could pick either a green or a yellow star.

Lesson 3: **Investigating chance**

- Investigate chance and record the results

Let's learn

Statistics and Probability

Guided practice

A class investigated the chance of choosing a yellow or a green cube from this bag.

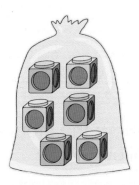

This tally chart shows the results. Use the tally chart to answer the questions.
How many?

	Tally
yellow	IIII IIII
green	IIII III

a yellow 10

b green 8

Lesson 4: **Presenting and describing data**

Key words
- **chance**
- **pattern**
- **random**
- **regular**
- **tally chart**
- **bar chart**
- **pictogram**

- Present and describe data on a chart or graph

Let's learn

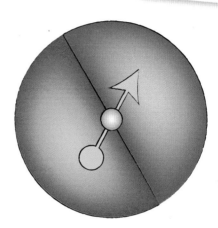

Number of spins — red, blue (bar chart)

Guided practice

There are 10 glasses of juice. 5 are orange and 5 are apple. 8 children are given a glass. Here are the results.

Juice

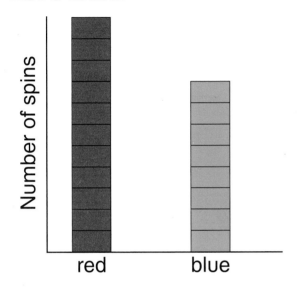

Juice	Number
orange	🥃🥃🥃🥃🥃
apple	🥃🥃🥃

You are given a glass from the drinks that are left. Do you think you will get orange or apple?

orange   (apple)

Why?

Because all of the orange has already been taken so only apple is left.

Statistics and Probability

113

The **Thinking and Working Mathematically Star**

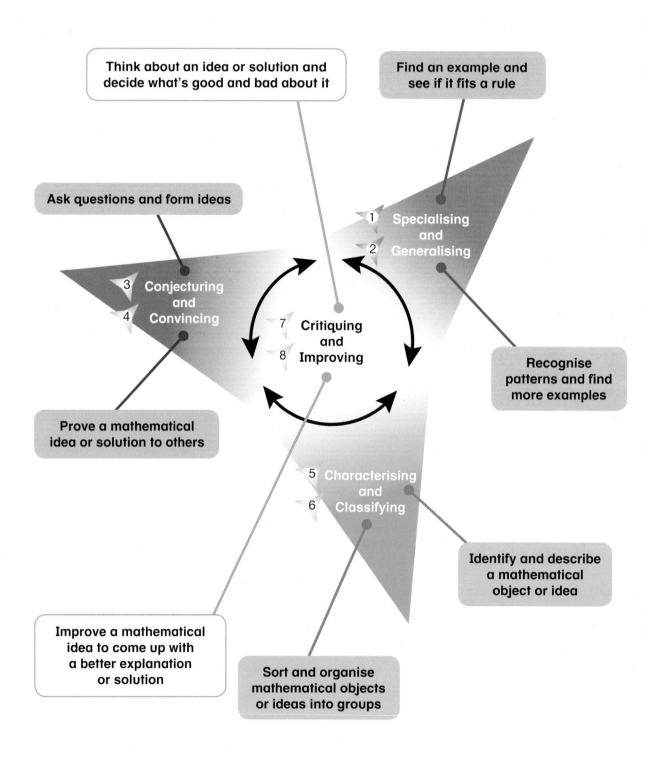

Think about an idea or solution and decide what's good and bad about it

Find an example and see if it fits a rule

Ask questions and form ideas

1
2 Specialising and Generalising

3
4 Conjecturing and Convincing

7
8 Critiquing and Improving

Recognise patterns and find more examples

Prove a mathematical idea or solution to others

5 Characterising and Classifying
6

Identify and describe a mathematical object or idea

Improve a mathematical idea to come up with a better explanation or solution

Sort and organise mathematical objects or ideas into groups

Acknowledgements

Photo acknowledgements

Every effort has been made to trace copyright holders. Any omission will be rectified at the first opportunity.

p6t Sirastock/Shutterstock; p18t LightField Studios/Shutterstock; p38c Superheang168/Shutterstock; p38b Ponysaurus/Shutterstock; p41t Yeamake/Shutterstock; p41b Kibri_ho/Shutterstock; p42t Koya979/Shutterstock; p43t Belander/Shutterstock; p45l Mahony/Shutterstock; p45c Michele Paccione/Shutterstock; p45r Vitaly Korovin/Shutterstock; p46b KittyVector/Shutterstock; p56l Ruth Black/Shutterstock; p56cll Ruth Black/Shutterstock; p56cl 5 second Studio/Shutterstock; p56c Ruth Black/Shutterstock; p56cr Ruth Black/Shutterstock; p56crr Jfunk/Shutterstock; p56r Africa Studio/Shutterstock; p58l Miniyama/Shutterstock; p73t Ohishiapply/Shutterstock; p76tl Sirtravelalot/Shutterstock; p77t Lilu330/Shutterstock; p78tl Mattasbestos/Shutterstock; p78tr Stockshoppe/Shutterstock; p78bl Ramziya Khusnullina/Shutterstock; p78br A_Ple/Shutterstock; p80b Szefei/Shutterstock; p81tl Szefei/Shutterstock; p81tr Vectorforjoy/Shutterstock; p81bl Szefei/Shutterstock; p81br Maxfromhell/Shutterstock; p83tl TopLevel/Shutterstock; p83tc Irin-k/Shutterstock; p83tr Jonathan Lewis/Shutterstock; p83cl Notbad/Shutterstock; p83bl Oleksandr Derevianko/Shutterstock; p83bc Mayakova/Shutterstock; p83br FARBAI/Shutterstock; p84t Artnata/Shutterstock; p85t Stock2You/Shutterstock; p86tl Galichstudio/Shutterstock; p86tr Jason Salmon/Shutterstock; p86cl Pixel-Shot/Shutterstock; p86c Tribalium/Shutterstock; p86cr Pixel-shot/Shutterstock; p86b 5 second Studio/Shutterstock; p91tl Ivn3da/Shutterstock; p91tr VectorUpStudio/Shutterstock; p91bl Zizi_mentos/Shutterstock; p91bc GraphicsRF/Shutterstock; p91br VladisChern/Shutterstock; p93tl Smart vision Baku/Shutterstock; p93tr Olga Danylenko/Shutterstock; p93b Hayati Kayhan/Shutterstock; p95bl Sashkin/Shutterstock; p95bcl Cobalt88/Shutterstock; p95bcr Macondo/Shutterstock; p95br Kibri_ho/Shutterstock; p97t Picturepartners/Shutterstock; p98bl Hchjjl/Shutterstock; p99tl MoonRock/Shutterstock; p99tr Top Vector Studio/Shutterstock; p100bl Bessyana/Shutterstock; p100bc Nikiteev_konstantin/Shutterstock; p100br 123Done/Shutterstock; p101t ESB Professional/Shutterstock; p102t Zaneta Baranowska/Shutterstock; p103b Dean Drobot/Shutterstock; p106t Monkey Business Images/Shutterstock; p107b Rvector/Shutterstock; p108l JIMMOYHT/Shutterstock; p108c Jemastock/Shutterstock; p108r Tinkivinki/Shutterstock; p113t Alex Staroseltsev/Shutterstock; p113b Den781/Shutterstock.